At Home
in the
Ocean
by Chris Layne

Contents

Science Vocabulary

Earth

Earth is the planet on which we live.

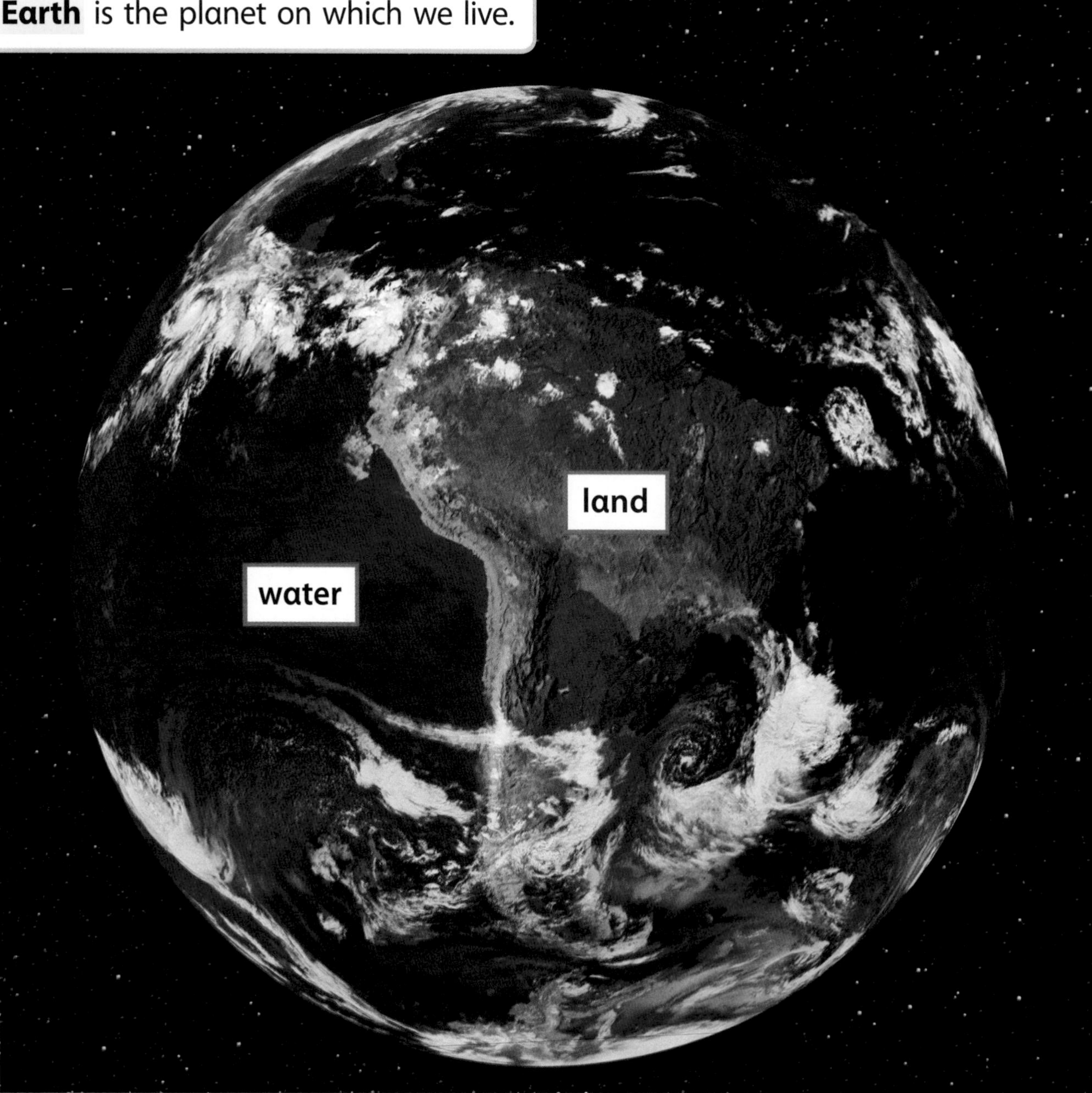

Earth has large areas of land and water.

habitat

A **habitat** is a place where living things can get what they need to stay alive.

The ocean is a **habitat.**

survive

When living things **survive,** they get what they need to stay alive.

A whale must have water to **survive.**

oxygen

Oxygen is a gas in air and water.

This fish has gills to take in **oxygen** from water.

shelter

A **shelter** is a safe place where a living thing can make its home and grow.

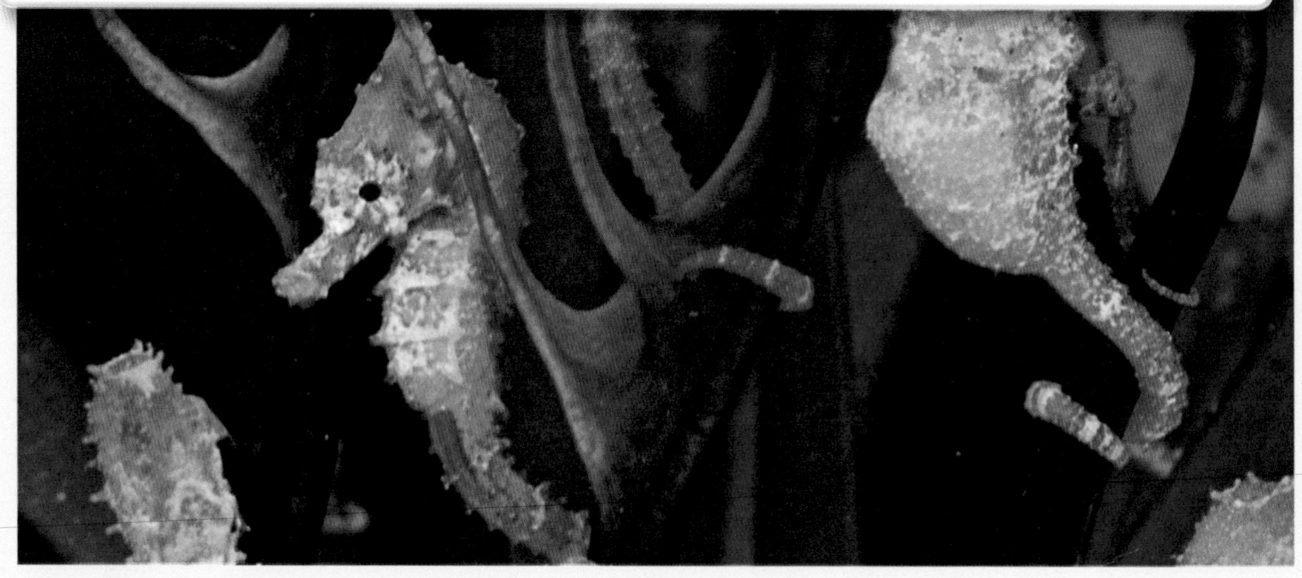

This sea grass is a **shelter** for the seahorses.

nutrients

Nutrients are parts of food and soil. They help living things stay healthy and grow.

This sea otter gets **nutrients** when it eats a crab.

My Science Vocabulary

- Earth
- energy
- habitat
- nutrients
- oxygen
- shelter
- survive

energy

Energy is the ability to do active things.

The sea grass gets **energy** from the sun. The green sea turtle gets energy when it eats the sea grass.

What Is the Ocean?

The ocean is a big body of salt water. It covers most of **Earth**.

Earth

Earth is the planet on which we live.

The ocean is wide. Parts are also deep!

The ocean is not deep near land. So the sun lights water near the surface of the ocean.

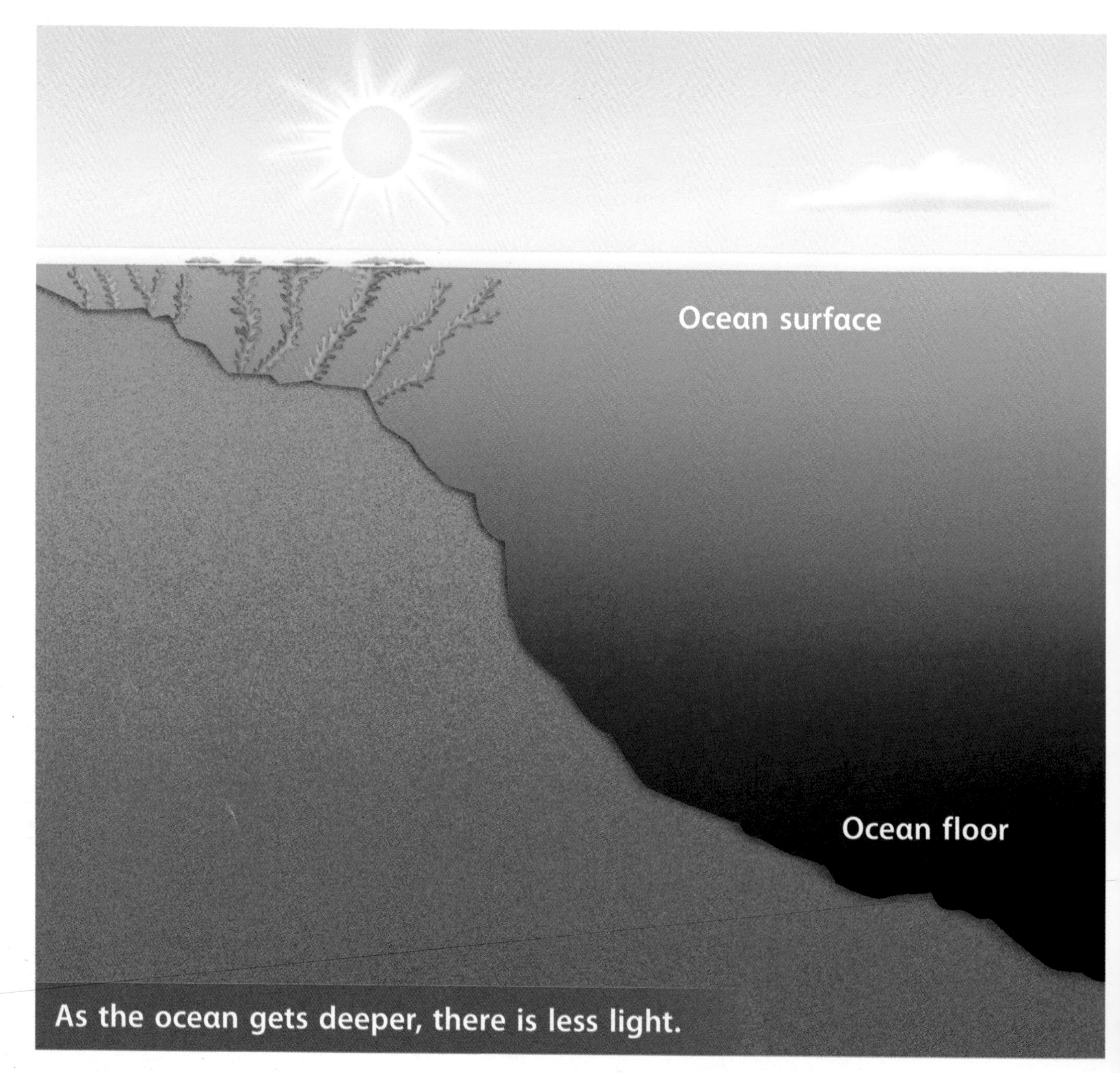

As the ocean gets deeper, there is less light.

The ocean has many levels. How are the levels different?

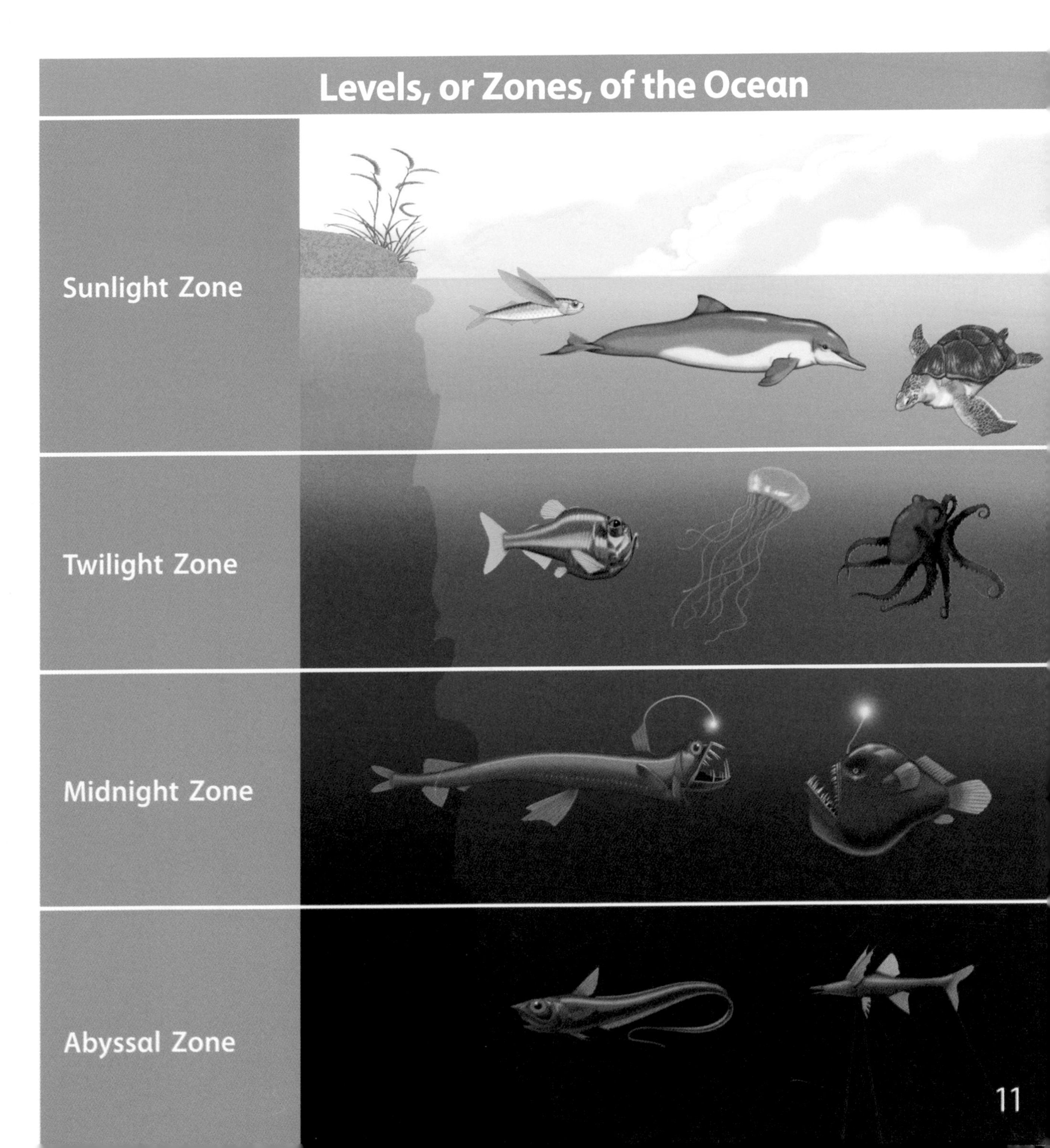

Each level of the ocean has a **habitat** full of life. Plants and animals live in each of these habitats.

habitat

A **habitat** is a place where living things can get what they need to stay alive.

Plants and animals have special parts. Special parts allow them to live in the ocean.

What special parts help these animals stay alive?

What Plants and Animals Need

All animals take in **oxygen.** They must have oxygen to **survive.**

Sea grass gives off oxygen that ocean animals need.

oxygen

Oxygen is a gas in air and water.

survive

When living things **survive,** they get what they need to stay alive.

Ocean animals get oxygen in different ways. How do these ocean animals get oxygen?

Fish have gills to take in oxygen from water.

Ocean mammals have lungs. They come to the surface to breathe air.

Animals also find food and **shelter** in plants. Plants help ocean animals survive. Some animals eat sea grass. Some animals find shelter in sea grass.

Sea grass is a plant that grows in shallow water near an ocean's shore.

shelter

A **shelter** is a safe place where a living thing can make its home and grow.

Animals that live in or near sea grass beds produce waste. This waste helps make the soil healthy. Sea grass needs the soil to grow.

Seahorses find shelter in sea grass.

Manatees eat sea grass.

Many ocean animals move to survive. They swim, float, or walk.

Animals called plankton drift with the current.

Lobsters walk on the ocean floor.

Some ocean animals don't move. They stay still to survive. Coral is an ocean animal that does not move.

Fish swim with the current over a coral reef.

Animals can move to catch food. They can also move to keep from being eaten!

A killer whale swims to catch fish for food. The fish swim away from the killer whale to keep from being eaten.

Some animals eat other animals for food. They get the **nutrients** they must have from the food they eat.

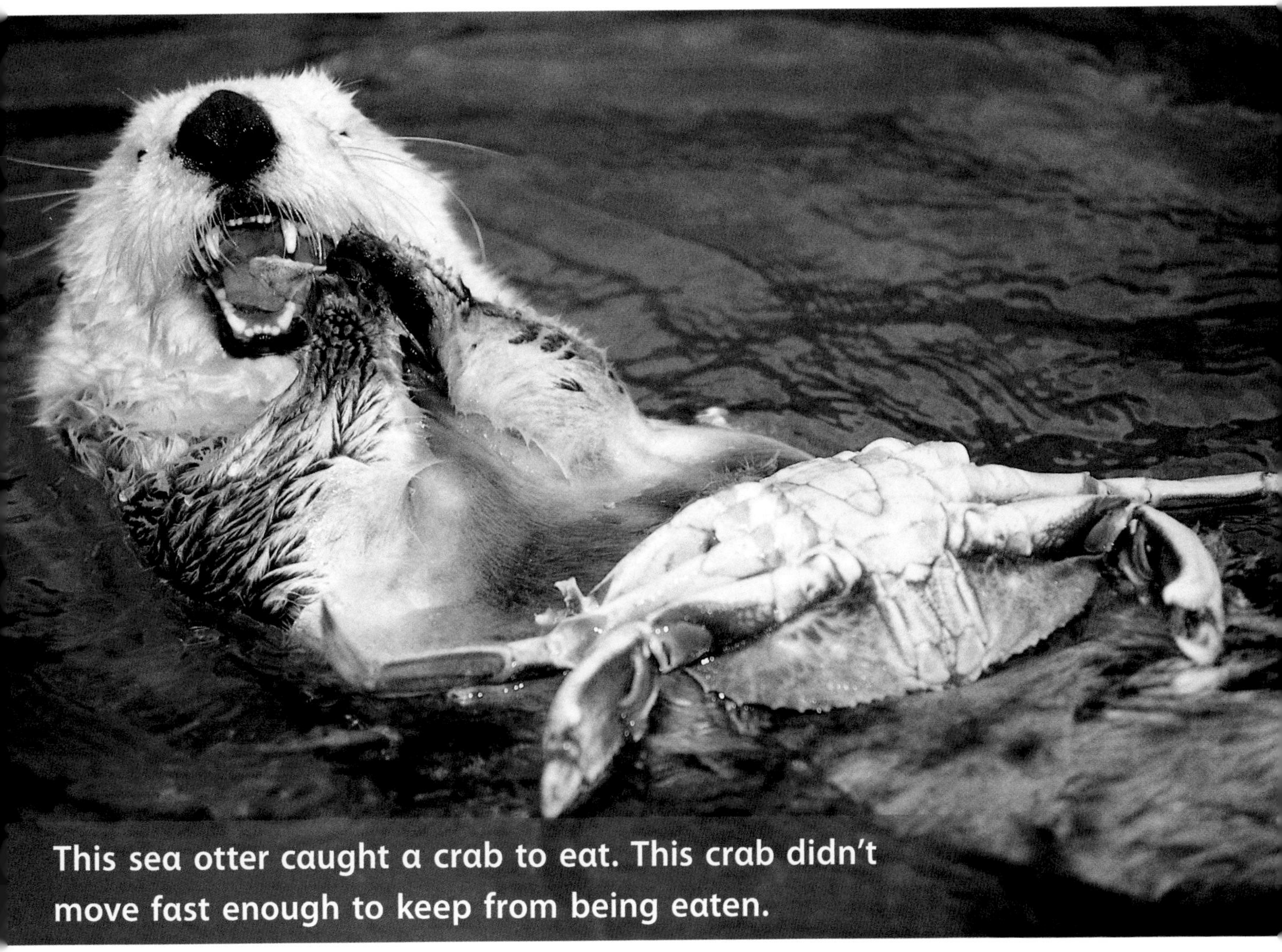

This sea otter caught a crab to eat. This crab didn't move fast enough to keep from being eaten.

nutrients

Nutrients are parts of food and soil. They help living things stay healthy and grow.

Many plants live in the ocean. Many animals do, too. Plants get **energy** from the sun.

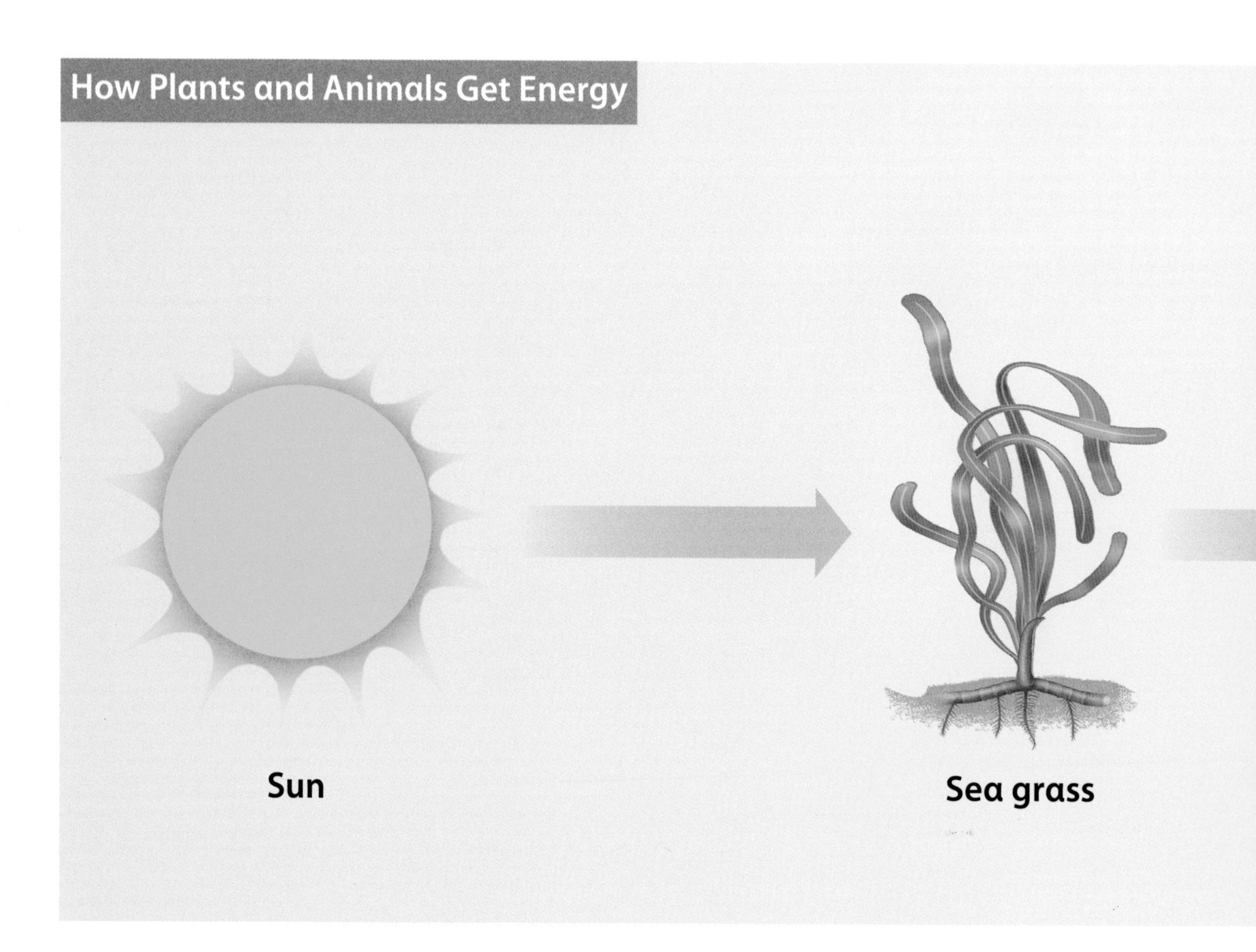

energy

Energy is the ability to do active things.

Animals get energy from eating plants and other animals. Sea grass is food for the green sea turtle. The green sea turtle is food for the shark.

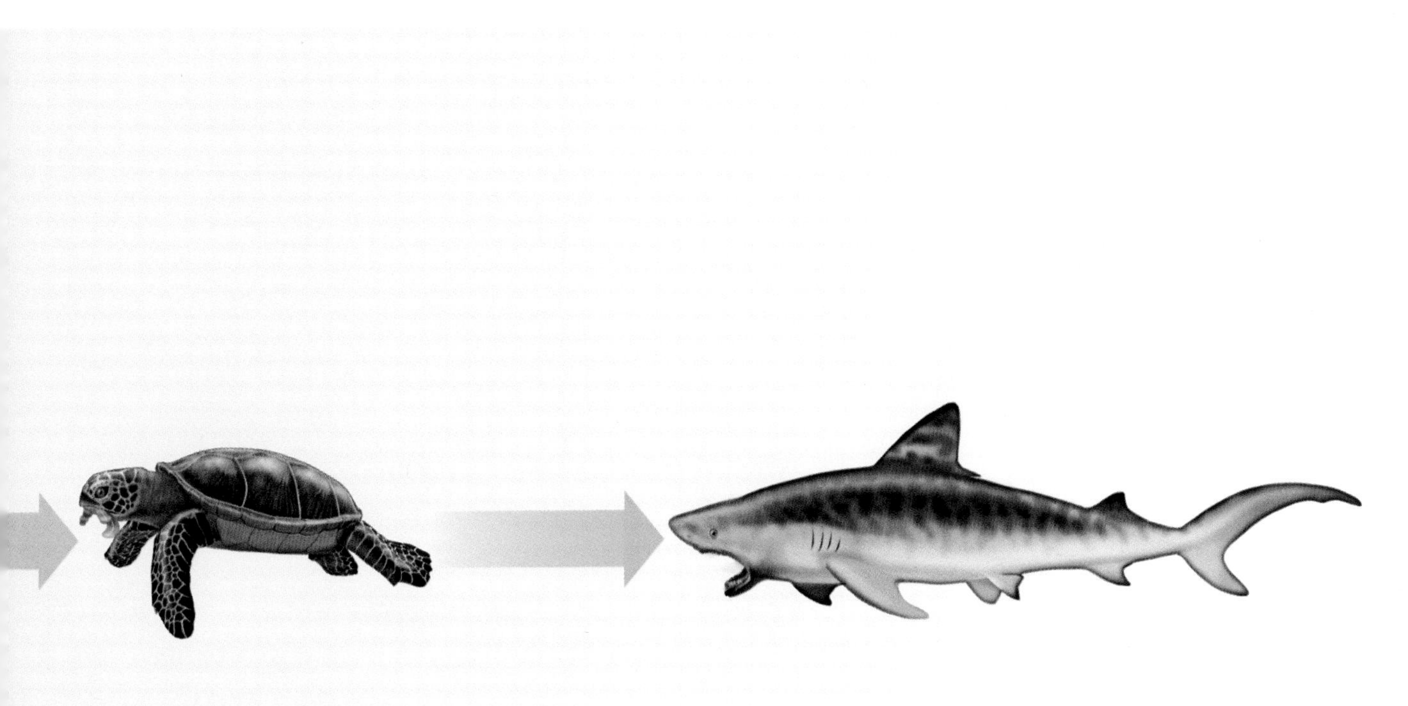

Green sea turtle

Shark

Ocean Habitats

Oceans cover most of Earth's surface.

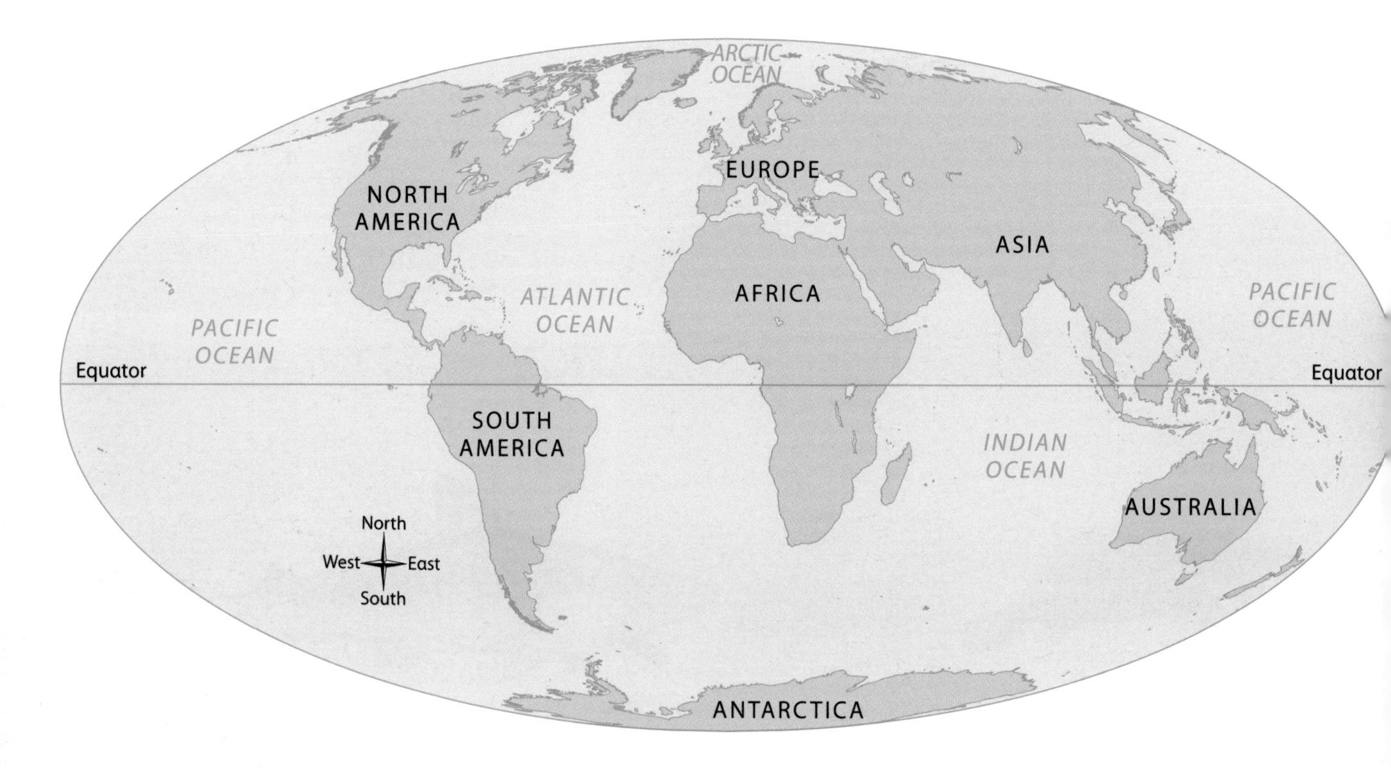

The ocean is one large body of water that has different regions. Each region has a name.

People get many things from the ocean. Some of our food comes from the ocean.

A fisherman uses a net to catch fish.

The ocean is filled with life. Yet there is more to learn about the ocean.

The ocean is a habitat for plants, fish, and other animals.

So scientists are exploring it. They are learning new things and finding new plants and animals.

Scientists use submarines to explore the ocean.

Conclusion

Many plants and animals live in the ocean. They get what they need to survive in their ocean habitat. Many of these ocean plants and animals can't live in other habitats.

Think About the Big Ideas

1. Where do ocean plants and animals live?
2. What do ocean plants and animals need to survive?
3. How do ocean plants and animals depend on each other?

Share and Compare

Turn and Talk

Compare the habitats in your books. How are they different? How are they alike?

Read

Find your favorite part of the book and read it to a classmate.

Write

Tell what plants and animals need to survive. Share what you wrote with a classmate.

Draw

Show an animal or plant in its habitat. Share your drawing with a classmate.

Meet Greg Marshall

Scientists ask a lot of questions. Sometimes technology can help them answer these questions.

Greg Marshall wanted to see animals in their habitats. So he invented a camera and put it on different kinds of animals.

It took him many tries to get the camera exactly right. He called his invention *Crittercam*.

2000
2008

Index

Acknowledgments
Grateful acknowledgment is given to the authors, artists, photographers, museums, publishers, and agents for permission to reprint copyrighted material. Every effort has been made to secure the appropriate permission. If any omissions have been made or if corrections are required, please contact the Publisher.

Photographic Credits
Cover (bg) Digital Vision/Getty Images; Cvr Flap (t), 5 (t), 8-9 Steve Arnold/Shutterstock; Cvr Flap (c), 7 (t), 21, 28 Jeff Foott/Getty Images; Cvr Flap (b), 16 Jason Edwards/National Geographic Image Collection; Title (bg) Norbert Wu/Minden Pictures/National Geographic Image Collection; 2-3 Creatas/Jupiterimages; 4 NASA/Corbis; 5 (b), 15 (b) Corel; 6 (t), 15 (t) Corel; 6 (b), 17 (l) Paul Zahl/National Geographic Image Collection; 14 Martin Strmiska/Alamy Images; 17 (r) Corel; 18 (t) Jean-Paul Ferrero/Minden Pictures, (b) Federico Cabello/SuperStock; 19 Digital Vision/Getty Images; 20 Amos Nachoum/Corbis; 25 (bg) Randy Olson/National Geographic Image Collection, (inset) Chris McLennan/Alamy Images; 26 DigitalStock/Corbis; 27 Stuart Westmorland/Science Faction/Getty Images; 30-31 (bg) Annie Griffiths Belt/National Geographic Image Collection; 30-31 (insets) Greg Marshall/ National Geographic Remote Imaging; 31 Peter McBride; Inside Back Cover (bg) Digital Vision/Getty Images.

Illustrator Credits
7 (l,c) John Kurtz; 7 (r) Paul Mirocha; 11 John Kurtz; 12-13 Rose Berlin/Cornell & McCarthy; 22 John Kurtz; 23 Paul Mirocha; 24 Mapping Specialists.

Program Authors
Randy Bell, Ph.D., Associate Professor of Science Education, University of Virginia, Charlottesville, Virginia; Malcolm B. Butler, Ph.D., Associate Professor of Science Education, University of South Florida, St. Petersburg, Florida; Kathy Cabe Trundle, Ph.D., Associate Professor of Early Childhood Science Education, The Ohio State University, Columbus, Ohio; Nell K. Duke, Ed.D., Co-Director of the Literacy Achievement Research Center and Professor of Teacher Education and Educational Psychology, Michigan State University, East Lansing, Michigan; Judith Sweeney Lederman, Ph.D., Director of Teacher Education and Associate Professor of Science Education, Department of Mathematics and Science Education, Illinois Institute of Technology, Chicago, Illinois; David W. Moore, Ph.D., Professor of Education, College of Teacher Education and Leadership, Arizona State University, Tempe, Arizona

National Geographic School Publishing
Hampton-Brown
www.NGSP.com

Printed in the USA.
RR Donnelley, Johnson City, TN

ISBN: 978-0-7362-5369-7

10 11 12 13 14 15 16 17

10 9 8 7 6 5 4 3 2